Uncle Steve, may I eat on the deck?

Craig settles on a bench. The sun dazzles him.

Craig rumples the hair on Bandit's back.

Craig gives Bandit a little sample of toast.

Brindle cuddles. She rumbles and gets a little toast.

A sparrow flutters. Craig sprinkles some toast.

A chipmunk wants a nibble.

If it is not a hassle, may I have some toast?

Can you use a tape measure?

I measure the room.

I measure the doorway.

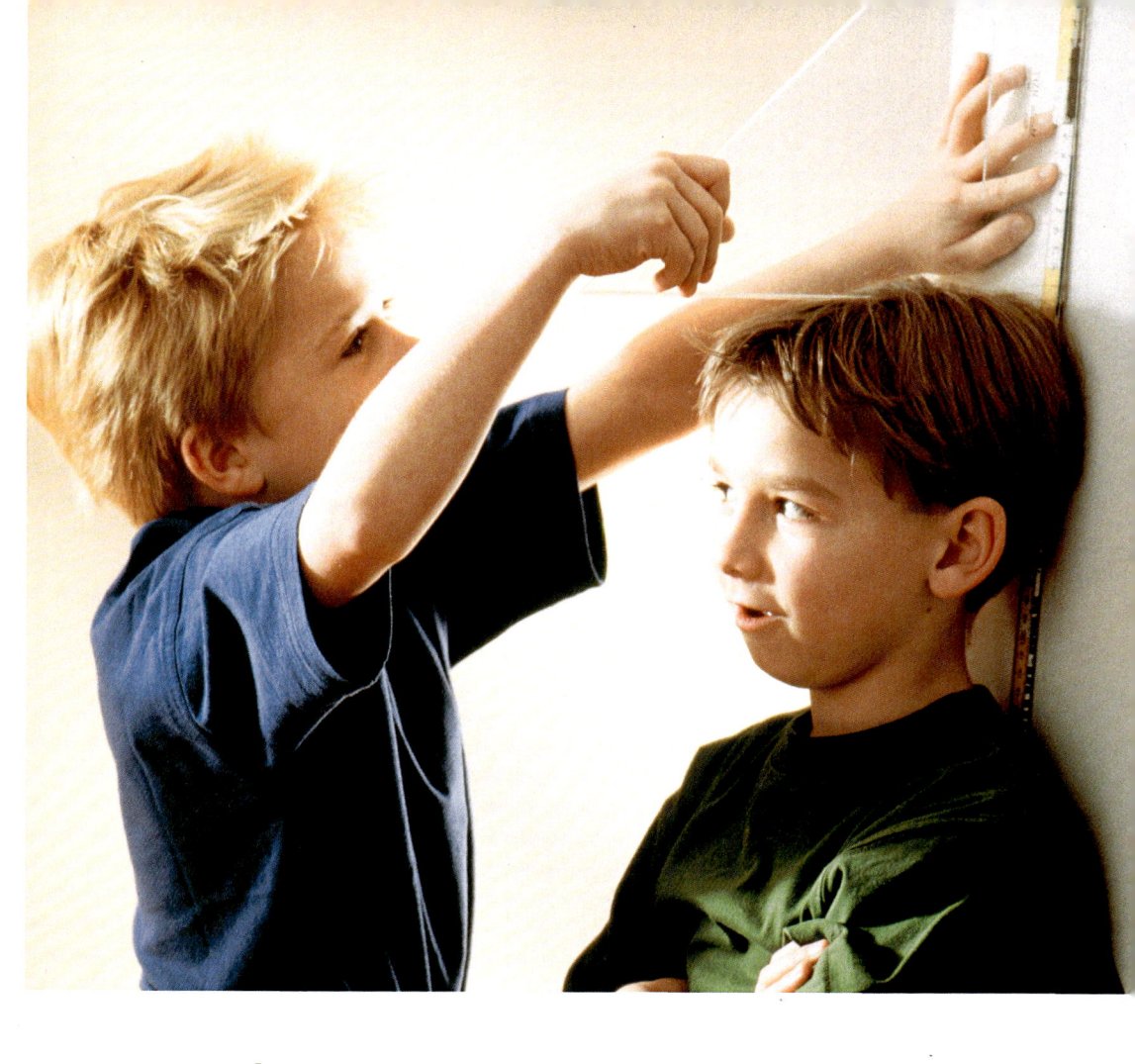

I measure my friend.

I measure the table.

I measure length.

What do you do with a tape measure?

Tape Measure

by Sheila Rivera

Lerner Publications · Minneapolis